P9-EDW-066

THE GULF OF MEXICO

THE GULF OF MEXICO

Text by Bern Keating
Photographs by Dan Guravich

A Studio Book
THE VIKING PRESS
New York

First published in 1972 by The Viking Press, Inc.
625 Madison Avenue, New York, N.Y. 10022
in association with Exxon Corporation

EXXON

Published simultaneously in Canada by
The Macmillan Company of Canada Limited
SBN 670-35759-6
Library of Congress catalog card number: 72-81678
Printed in U.S.A.

ACKNOWLEDGMENT

The publishers wish to thank Lawrence F. Mihlon and William L. Copithorne for their valuable assistance in the conception and development of the Exxon Shorelines of America series.

CONTENTS

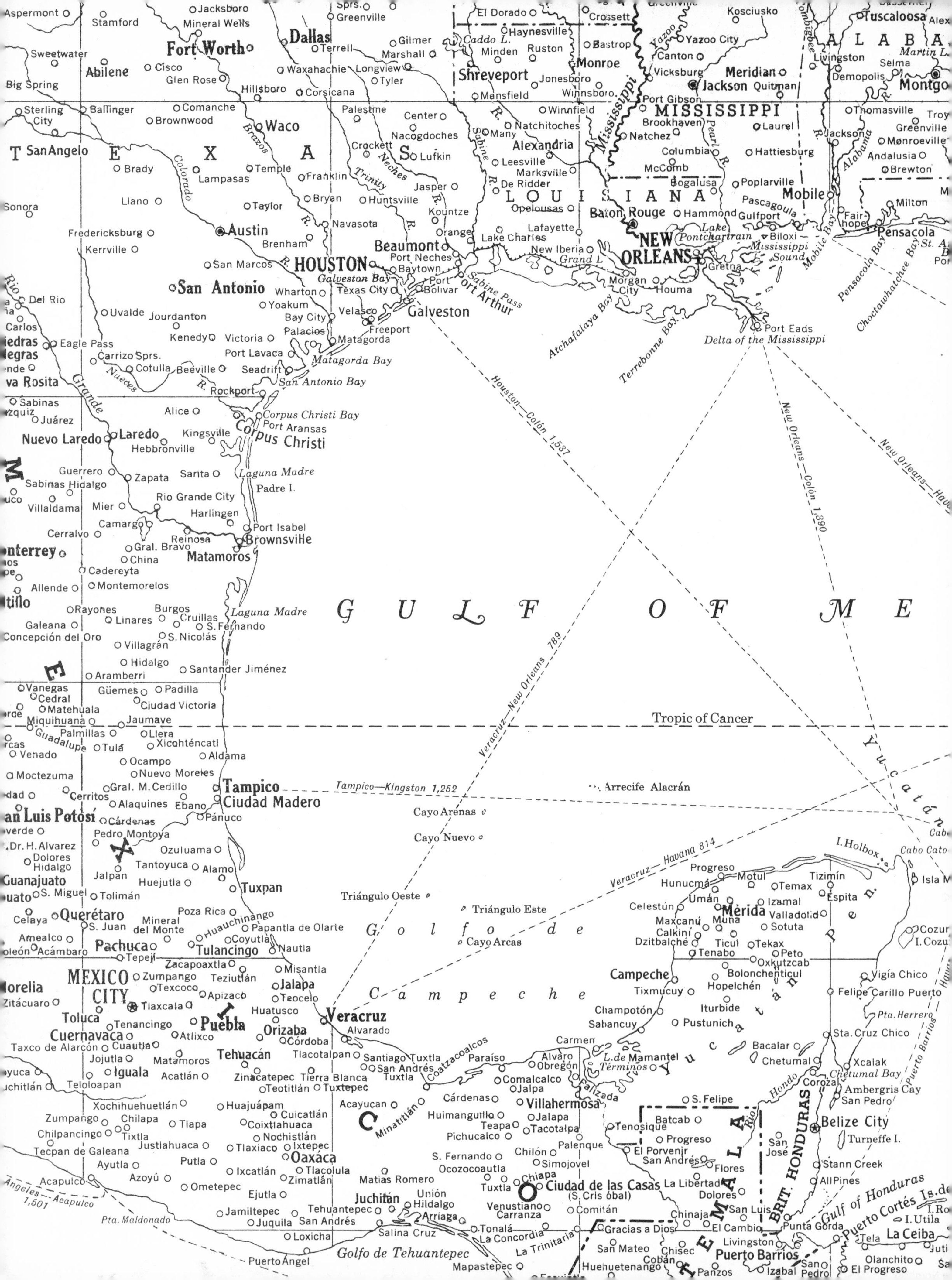

GULF OF ME
Golfo de Campeche
Tropic of Cancer
T E X A S
LOUISIANA
MISSISSIPPI
ALABA
M E X
Yucatán Pen.
Aspermont
Stamford
Jacksboro
Mineral Wells
Greenville
El Dorado
Crossett
Kosciusko
Tuscaloosa
Haynesville
Caddo L.
Yazoo
Yazoo City
Sweetwater
Fort Worth
Dallas
Terrell
Gilmer
Marshall
Minden
Ruston
Bastrop
Canton
Livingston
Selma
Martin L.
Abilene
Cisco
Waxahachie
Longview
Shreveport
Monroe
Vicksburg
Meridian
Demopolis
Montgo
Big Spring
Glen Rose
Tyler
Jonesboro
Jackson
Quitman
Hillsboro
Corsicana
Mansfield
Winnsboro
Port Gibson
Sterling City
Ballinger
Comanche
Palestine
Center
Winnfield
Thomasville
Troy
Brownwood
Waco
Natchitoches
Brookhaven
Laurel
Greenville
Crockett
Nacogdoches
Many
Natchez
Jackson
Monroeville
San Angelo
Lufkin
Alexandria
Columbia
Hattiesburg
Andalusia
Brady
Temple
Leesville
Marksville
McComb
Brewton
Lampasas
Franklin
Jasper
De Ridder
Bogalusa
Poplarville
Mobile
Sonora
Llano
Taylor
Bryan
Huntsville
Opelousas
Baton Rouge
Hammond
Pascagoula
Gulfport
Milton
Navasota
Kountze
Lafayette
Lake Pontchartrain
Biloxi
Fairhope
Pensacola
Fredericksburg
Austin
Orange
Lake Charles
Kerrville
Brenham
Beaumont
Port Neches
New Iberia
NEW ORLEANS
Mississippi Sound
San Marcos
HOUSTON
Baytown
Grand L.
Gretna
Pensacola Bay
Choctawhatchee Bay
Galveston Bay
Port Bolivar
Port Arthur
Sabine Pass
Morgan City
Houma
San Antonio
Wharton
Texas City
Del Rio
Yoakum
Galveston
Uvalde
Jourdanton
Bay City
Velasco
Freeport
Atchafalaya Bay
Terrebonne Bay
Port Eads
Delta of the Mississippi
Eagle Pass
Kenedy
Victoria
Palacios
Matagorda
Carrizo Sprs.
Port Lavaca
Matagorda Bay
Cotulla
Beeville
Seadrift
San Antonio Bay
Rockport
Sabinas
Alice
Corpus Christi Bay
Juárez
Port Aransas
Nuevo Laredo
Laredo
Kingsville
Corpus Christi
Hebbronville
Houston—Colón 1,537
New Orleans—Colón 1,390
New Orleans—Hava
Guerrero
Sabinas Hidalgo
Zapata
Sarita
Laguna Madre
Padre I.
Rio Grande City
Villaldama
Mier
Harlingen
Camargo
Port Isabel
Cerralvo
Reinosa
Brownsville
Gral. Bravo
China
Matamoros
Cadereyta
Allende
Montemorelos
Rayones
Burgos
Cruillas
Laguna Madre
Galeana
Linares
S. Fernando
Concepción del Oro
S. Nicolás
Villagrán
Hidalgo
Aramberri
Santander Jiménez
Vanegas
Güemes
Padilla
Cedral
Matehuala
Ciudad Victoria
Miquihuana
Jaumave
Palmillas
Llera
Tula
Xicohténcatl
Venado
Aldama
Ocampo
Moctezuma
Nuevo Morelos
Gral. M. Cedillo
Tampico
Tampico—Kingston 1,252
Arrecife Alacrán
Cerritos
Alaquines
Ebano
Ciudad Madero
San Luis Potosí
Cárdenas
Pánuco
Cayo Arenas
Pedro Montoya
Cayo Nuevo
Dr. H. Alvarez
Ozuluama
Dolores Hidalgo
Tantoyuca
Alamo
Veracruz—New Orleans 789
Veracruz—Havana 814
I. Holbox
Cabo Catoche
Progreso
Guanajuato
Jalpan
Huejutla
Tuxpan
Motul
Tizimín
Isla Mujeres
S. Miguel
Tolimán
Triángulo Oeste
Hunucmá
Temax
Espita
Querétaro
Poza Rica
Triángulo Este
Celestún
Umán
Izamal
Celaya
S. Juan
Mineral del Monte
Huauchinango
Papantla de Olarte
Mérida
Maxcanú
Muna
Valladolid
Amealco
Calkiní
Sotuta
Cozumel
Acámbaro
Pachuca
Coyutla
Cayo Arcas
Dzitbalché
Ticul
Tekax
Tepeji
Tulancingo
Nautla
Tenabo
Peto
Zacapoaxtla
Misantla
Oxkutzcab
Morelia
MEXICO CITY
Zumpango
Teziutlán
Jalapa
Campeche
Bolonchenticul
Vigía Chico
Texcoco
Hopelchén
Tixmucuy
Felipe Carillo Puerto
Zitácuaro
Tlaxcala
Apizaco
Teocelo
Toluca
Huatusco
Veracruz
Champotón
Iturbide
Pta. Herrero
Tenancingo
Puebla
Orizaba
Sabancuy
Pustunich
Cuernavaca
Atlixco
Alvarado
Córdoba
Sta. Cruz Chico
Taxco de Alarcón
Cuautla
Carmen
Bacalar
Jojutla
Matamoros
Tehuacán
Tlacotalpan
Santiago Tuxtla
Coatzacoalcos
Paraíso
Alvaro Obregón
L. de Mamantel
Términos
Chetumal
Xcalak
Iguala
Acatlán
San Andrés Tuxtla
Comalcalco
Chetumal Bay
Teloloapan
Zinacatepec
Tierra Blanca
Jalpa
Corozal
Ambergris Cay
Teotitlán
Tuxtepec
Palizada
Hondo
Puerto Barrios
Xochihuehuetlán
Cárdenas
Villahermosa
S. Felipe
San Pedro
Huajuápam
Acayucan
Minatitlán
Huimanguillo
Zumpango
Chilapa
Tlapa
Cuicatlán
Teapa
Jalapa
Tenosique
Batcab
Coixtlahuaca
Tacotalpa
Belize City
Chilpancingo
Tixtla
Nochistlán
Pichucalco
Progreso
Turneffe I.
Justlahuaca
Ixtepec
Palenque
San José
Tecpan de Galeana
Tlaxiaco
S. Fernando
Chilón
El Porvenir
Oaxáca
Simojovel
San Andrés
Flores
Stann Creek
Ayutla
Putla
Ocozocoautla
Azoyú
Ixcatlán
Tlacolula
Chiapa
AllPines
Acapulco
Zimatlán
Matías Romero
Tuxtla
Ciudad de las Casas
La Libertad
Gulf of Honduras
Ometepec
Ejutla
Juchitán
Unión Hidalgo
(S. Cristóbal)
Dolores
Puerto Cortés
Angeles—Acapulco 1,501
Jamiltepec
Tehuantepec
Venustiano Carranza
Comitán
San Luis
BRIT. HONDURAS
Juquila
San Andrés
Arriaga
Chinaja
Punta Gorda
I. Utila
Pta. Maldonado
Loxicha
Salina Cruz
Tonalá
La Concordia
La Trinitaria
Gracias a Dios
El Cambio
Tela
La Ceiba
Golfo de Tehuantepec
Livingston
Puerto Angel
Mapastepec
San Mateo
Chisec
Cobán
Huehuetenango
Panzos
Izabal
San Pedro
Olanchito
El Progreso

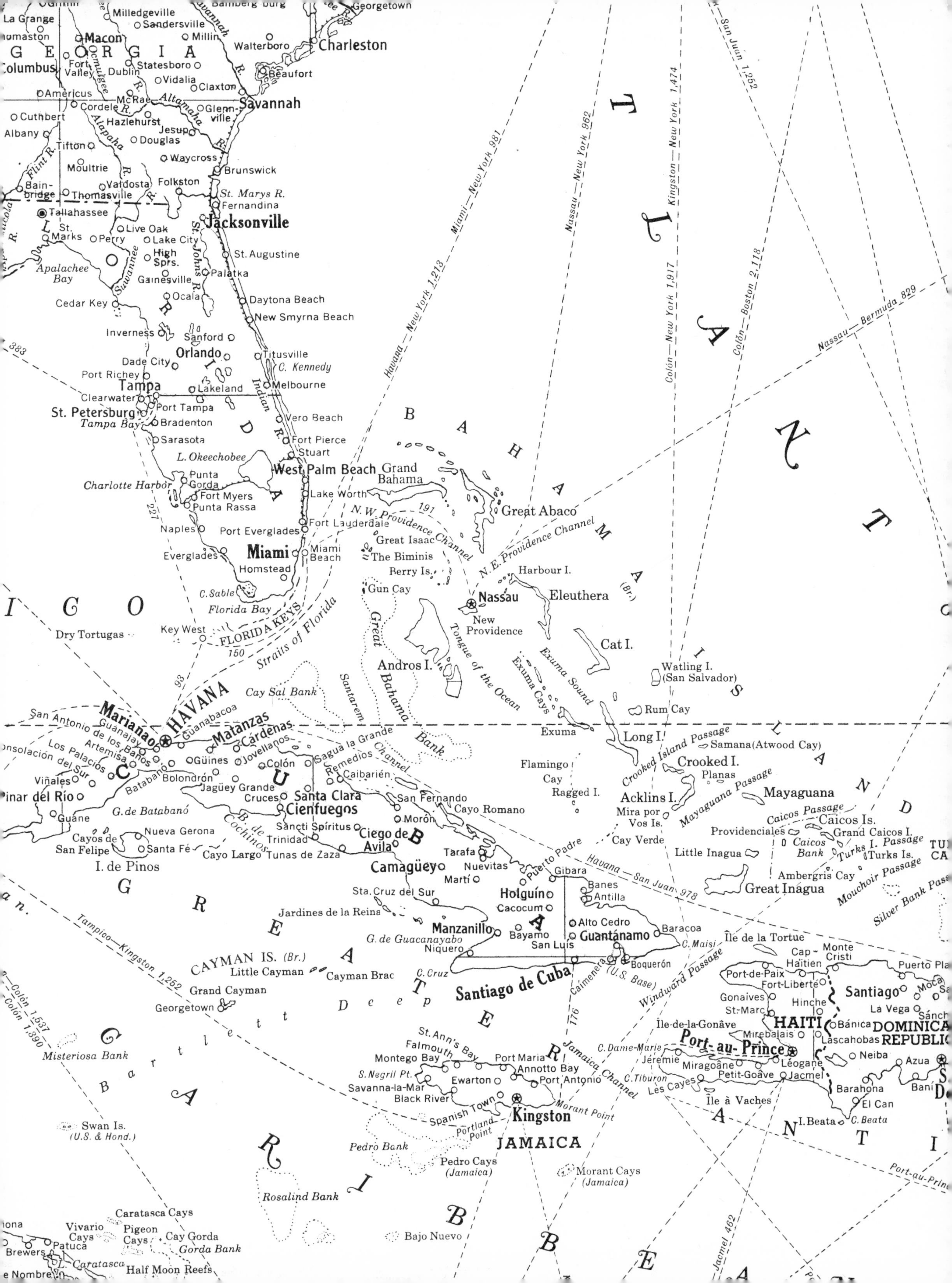

How Much Lost: How Much Left?

Our patrol boat slammed across wave troughs, dug halfway through oncoming seas, and plunged with thrashing propellers over wave crests, throwing sheets of spray in dazzling arcs. Jarred so hard while sitting that I was afraid of breaking a tooth, I stood and clung with aching knuckles to a handrail for the six-hour trip from Key West to the farthest outlying dot of American territory in the Gulf of Mexico at the sun-blasted coral keys called the Dry Tortugas. There I was to meet a plane and pilot to begin a survey of the Gulf Shore.

We had run by the bird-swarming but uninhabited Marquesas Keys and over the shallow Quicksands, strewn with shipwrecks once used as naval targets and now rotting on the bottom, across Halfmoon Shoals where the hulk of a sunken Cuban ferry wavered dimly visible just beneath our keel, and across stormy Rebecca Shoals. Above the roar of the wind and the booming of our passage through heaving seas, the radio popped and crackled a running account of another struggle in the centuries-old squabble over the marine riches of the Gulf.

Among dozens of trawlers scattered about the horizon, harvesting the stupendous schools of shrimp that frequent shallow seas about the Tortugas, a Coast Guard ship had jumped a Cuban vessel dragging American territorial waters. Through salt-fogged binoculars, I watched Coast Guardsmen cut the poacher out from the pack and shepherd it toward a mainland showdown.

From our boat's wheel, Patrolman Jim McCarthy flashed a grin out of a young face already burned a permanent mahogany-brown by tropical sun reflected from Gulf waters.

"Me and my buddies caught us a mess of Cuban crawfishers a couple months ago. So many I couldn't use my .38 pistol to persuade them I was a proper man. Had to point a Thompson gun to make them believe in me.

"But so long as shrimps and crawfish bring the price they do and they go on swarming around these keys, you ain't gone keep all them poachers out. No way. People been fighting over the Gulf since long before us white folks got into the act. You know Key West means 'Bone Island' in Spanish 'cause the place was covered with skeletons of Indians who had busted each other's heads open arguing over who was gonna eat the fish fry. And them Indians was kindly compared to the folks come to the Gulf behind them."

And truly the shores of the Gulf of Mexico did lure the first white explorers of continental North America, the ruthless Conquistadors come to fatten on the riches actually or supposedly hidden around the fecund basin.

Before the first northern Europeans timidly tried to colonize the Atlantic seaboard, Spanish-speaking universities had graduated several generations in provinces served by the Gulf. When

Atlantic Coast colonists first sent to Europe niggardly shipments of an Indian weed the homefolk had to become addicted to before it represented wealth, the Gulf colonies had been sending home for half a century immense flotillas wallowing under overloads of gold and silver bullion so rich that they wrecked Europe's monetary system.

Today the Gulf and its shores ship to markets of the world even more immense riches, such as oil, sulphur, salt, cattle, furs, and fish. And Gulf Shore dwellers grow fat tending flocks of suntanned tourists that yield uncountable millions in golden fleece.

Almost one-quarter of the continent to the north feels the Gulf's moderating presence, for another export from that semi-landlocked warm water basin is weather. When low-pressure systems far inland suck moisture-laden air up from the Gulf, rains fall, spoiling garden weddings but making the cotton, corn, and sugar cane grow tall. Arctic cold air masses blown south of normal storm tracks almost always veer back northward when they come against the warm air mass sitting sluggish and immovable over the tepid Gulf. And it is no insignificant air mass, for it is heated by one of the world's great natural warming baths.

With roughly 600,000 square miles of surface, the Gulf of Mexico ranks fifth among the world's seas, surpassed only by the South China, Caribbean, Mediterranean, and Bering seas. Though the Gulf is only slightly more than half the size of the neighboring Caribbean, it is nevertheless half again as large as the combined five bordering states of Texas, Louisiana, Mississippi, Alabama, and Florida. Not counting bays and inlets, the coastline runs about 3000 miles, half of it in the United States, the other half in Mexico.

And that immense sea is almost landlocked. It is shaped like a squat flask with only a narrow mouth between the tips of the Florida and Yucatán peninsulas. Like a bottle stopper, the island of Cuba half corks the flask's already narrow neck. The Cuban cork allows the warm Yucatán Current to flow into the flask only through the 120-mile-wide channel at the southeastern lip of the opening. The Florida Current pours out through a 60-mile-wide channel between the Florida Keys and Cuba's north shore. That Florida Current, sometimes flowing faster than a man walks, becomes the Gulf Stream, a river 1000 times the size of the Mississippi, carrying its warm Gulf waters to the distant shores of Iceland and Northern Europe so that burghers living in what should be a subarctic climate can garden in shirt sleeves as early as April.

Long a lover of the soft climate and rich wildlife of the Gulf's shorelands, I had planned to circle it to see how its beauties were standing up to the ravages of progress, at least along the United States coast. First, I climbed aboard a plane in the Tortugas and began a low-level flight around the rim of the basin as far as the Rio Grande, making a quick search for strips of coast that merit closer inspection.

Thunderheads still marched in stately procession across the vastness of the Everglades, manatees still roiled muddy bottoms of the flats at Cape Romano, impenetrable swamps teeming with wildlife still brooded around Florida's Big Bend, magnificent white sand dunes lined the Panhandle shores, the great Louisiana saw grass marshes still stretched from the Mississippi's delta to the Sabine River, barrier islands of Texas reaching down to the Rio Grande still sheltered uncountable flocks of wintering birds, and lagoons between barrier and mainland floated rafts of waterfowl in astronomical numbers.

But vast stretches of the wild coastline had disappeared

under assault of bulldozer and dragline to make boatarama cities and mobile home retirement villages. Auto junkyards rimmed the cities. Palls of smoke hid thousands of acres around industrial sites. Perhaps from the air the destruction of what seemed to be left was invisible and the region was already destroyed.

Disembarking from the plane, I began my terrestrial survey with worried heart. The question was:

Is there enough natural beauty left around the Gulf of Mexico to make it worth fighting to save?

The Several Floridas

One winter day the norther blew into Pensacola on Florida's Panhandle coast. Pedestrians scuttled across wind-blasted downtown streets and ducked behind the steamy windows of corner cafés to warm their bones over a cup of coffee. That same day, five hundred miles to the southeast across Gulf waters, I snorkeled in the warm seas off Loggerhead Key, searching for the waving telltale antennae that mark a crawfish hiding place.

For around the Gulf Shore there are several Floridas with differing climates: the coral reefs and limestone outcroppings of the Keys, the mangrove swamps of the Ten Thousand Islands, the sand beaches of the Sun Coast, the tideless flats of the Big Bend, and the dunes of the Panhandle. Each zone offers its peculiar habitat to the sea and land creatures that find a home there. Each zone has its defenders—and its spoilers who are almost as exquisitely adapted to ruining the habitat as the wild creatures are adapted to living there.

Safest from the spoilers of all the habitats is the semidesert isolation of the Dry Tortugas, sixty-five miles across shoal-infested waters from Key West, for the isles lie under the protection of the National Park Service.

The grim inhospitality of the seven isles of the Dry Tortugas is more apparent than real; a few of the turtles that gave the islands their name still crawl the sand beaches to lay their eggs, the seas teem with marine life, and countless thousands of birds of dozens of species land there to break incredibly long migration flights across the pitiless Gulf.

The Audubon Society, conscious that the Tortugas are called "the Dry" for good reason, sank artesian wells in the parade ground of Fort Jefferson on Garden Key and at the lighthouse on Loggerhead Key. Within the fort where prisoners of the Civil War once languished, migrating birds slake their thirst at the constantly welling fountain. Seed eaters can rebuild strength if they are not too demanding. But birds depending on gross insects, most notably the African cattle egrets invading North America from South American way stations, let themselves be seduced by the cool artesian waters and stay close by the bubbling fountains till they sicken and die of starvation—all but a ruthless few who violate their natural appetites by gobbling enfeebled songbirds to restore their strength and then wing off toward their mainland goal which they mysteriously know lies northward.

During the spring tens of thousands of sooty terns arrive at Bush Key across the narrow channel from the crumbling brick walls of the fort. Probably the most numerous terns in the world, the sooties nevertheless breed nowhere in North America but on that one desolate coral head. After they raise their young, they leave, but nobody knows where they go. They cannot rest on the waves, for they quickly become waterlogged and sink. And nobody has found where all the millions of them roost when they are away from their nests.

On Bush Key they checkerboard the sands with invisible territorial boundaries marking off squares about eighteen inches to a side. Each pair of terns guards an egg in the center of their square. So ferociously do they defend their minuscule kingdom that they will slaughter even a fuzzy chick that wobbles across the frontier.

In 1565 an English slaver recorded in his log that he loaded his ship with thousands of seabirds nesting in the Tortugas—almost certainly the ancestors of these same sooty terns. John James Audubon in 1832 visited the rookery where the birds were "a cloud-like mass" and the sailors knocked down birds with sticks.

"In less than half an hour, more than a hundred terns lay dead in a heap and a number of baskets were filled to the brim with eggs."

Audubon met a party of Cuban egg hunters who had already collected "about eight tons of the eggs of the tern and the noddy."

Ravages of meat and egg hunters cut the tern population from what must have been close to a million in Audubon's day to a low of 3600 in 1907 when wardens began protecting the breeding grounds. In 1930 the National Park Service took over all the islands. (The original Bird Key disappeared in the hurricane of 1933, but the terns have moved their rookery to nearby Bush Key.)

In a rare happy note, I can report that on my visit to the island I had to walk a narrow beach strip, for I could not put a foot down in the interior without the danger of crushing an egg. Birds swirled about my ears in a dense cloud. I cannot estimate the population—how do you count a swarm of flying birds a mile or more in diameter?—but the sooty tern is at least 200,000 birds away from being an endangered species in North America, and may be many times more numerous than that.

The turtle story is not so happy. Once the nesting ground of thousands of turtles, the Dry Tortugas now shelter at most only a few hundred nests, and possible only a few score.

On the trip back to Key West with the Marine Patrol officers, we came across a pair of green turtles mating in the open sea. The species, once numerous, has become so rare that even those veteran sea rovers had never witnessed the single-minded coupling of those rapidly disappearing survivors of the Age of Reptiles. With great excitement they wheeled the patrol craft in high-speed zigzag chase of the turtles so I could take their picture. Diving and jinxing through underwater grass beds, the turtles never broke off their nuptials, a union that lasts from six to eight hours at a time, according to my guides.

"The old lady has to be the world's great navigator," McCarthy said. "An old loggerhead crawls up on a beach and lays her eggs one night and then disappears somewhere back in the sea. Three weeks later on a dark night she crawls back and lays another clutch of eggs within ten feet of the first nest.

"Now, I'm a pretty good sailor in these parts. I know these waters like I know the back of my hand. I got a seven-battery flashlight, a compass, a radio, all the charts I can carry. If you offered me a million dollars to plant a flag on a deserted beach one night and land within a hundred yards of that flag on a night three weeks later, I couldn't do it. How does that old turtle lady do it? It's enough to make you wonder."

The green turtle has condemned itself to extinction partly by furnishing man one of his most savory table delights. A bowl of green turtle soup, flavored with a dollop of amontillado, has few rivals in gastronomy.

Key West operates several crawls to hold green turtles brought up from the Cayman Islands until they are ready for the table. And the turtle is only one of the city's links with the Caribbean. Key West is more Caribbean than American—in vegetation, in cuisine, in outlook, and even in language, for Spanish is almost as common as English, and even the English of the native Conchs has an island flavor (as witness the way they pronounce the name in their own dialect so that it comes out *kahnk*). After all, the Caribbean Sea lies just across the narrow Florida Straits and the still narrower sliver of Cuba, itself a purely Caribbean land.

On a casual trip through the little city, I spotted sapodillas, tamarinds, Manila palms, frangipani, banyans, night-blooming cereus—and I could hardly miss the dazzling cascades of incandescent scarlet on blooming poinciana trees. For more than a century sailors have brought home seeds and cuttings from the tropical ports of the world, so Key West blooms with exotics like a botanical garden.

At a Cuban restaurant I dined on conch stew, crusty bread, an *alcaporado* of beef made of olives and raisins with fried plantains on the side and guava pudding for dessert. The menu was more Caribbean than any I've found in Jamaica.

From Key West you get to the mainland over a 135-mile-long highway that jumps from key to key over 42 bridges. Once it was the roadbed for a railroad, but the tracks were wiped out by the Labor Day hurricane of 1935 that drowned 425 Bonus Army veterans in a labor camp. The road curves northeasterly and parallels the Florida Current and Gulf Stream to the south. It also parallels a coral reef that the state's promoters insist on calling the only coral reef in North America—conveniently ignoring the magnificent Flower Gardens coral reef off the Gulf Coast near Galveston, Texas.

On the north side of the highway stretches Florida Bay, a lagoon so shallow that a strong wind can almost bare the bottom. When more of the earth's water was locked up in the vast glaciers of the last Ice Age, Florida was twice as big as it is now. The shallows of Florida Bay were drowned by melting ice only a few hundred generations ago, long after man had arrived on the scene. Divers find remains of prehistoric forests on the bay floor. Most of the bay is now in Everglades National Park, and rangers protect the rare species that survive there, including the great white heron that was almost exterminated in the disastrous 1935 hurricane but has made an encouraging recovery.

An even more spectacular success scored by conservationists is the comeback of the Key deer. The pygmy race of the Virginia whitetail deer of the mainland was hunted almost to extinction till a refuge sheltered them on Howe and Big Pine Keys. Down to perhaps 50 specimens in 1947, the deer now number more than 450 and have lost much of their timidity. At sunrise and sunset they openly browse on cottage lawns.

It's a heartwarming success story, but a weird and disturbing phenomenon goes with it. Game management, including controlled burns to improve forage, has so improved the range that the cute little deer are growing. They just might turn out to be nothing but big old ordinary whitetails like deer anywhere in eastern North America. So kindness may wipe out an apparent race that survived unchecked slaughter.

The semi-isolation of the Keys from the continent's biological mainstream, an isolation that made possible the development in only a few thousand years of a pygmy race of the common whitetail deer, for instance, makes of that steppingstone island world one of the most delightful homes for man this planet offers. Species of sea creatures, land plants, birds and animals unknown or rarely observed just a few miles north flourish in the Keys. And though the bulldozer operator works overtime destroying the beauty his clients have come to enjoy, many of the islets are so small or so unhandily located that they will probably escape development into the sun-baked horror of a pizza parlor parking lot. Many of them are merely clumps of red mangroves growing where a floating seed snagged on the bay bottom and took root.

Swiftly the mangrove spreads, seemingly walking across the water on its stiltlike roots, trapping silt and building a boggy little island, too small and wet for folks, but ideal for a host of birds and sea creatures seeking refuge from common enemies—including man.

Unfortunately, among the species that thrive under the shelter of the mangrove are the Key varieties of mosquito, as voracious as a horde of winged Genghis Khans slavering for blood.

These mosquitoes prompted an early resort developer in the Keys to build a bat tower and import a tribe of bats to control the pests. The bats promptly disappeared—devoured by the mosquitoes, according to old-timers.

When he travels through mangrove country like the Keys, even the most ardent defender of the environment has to keep reminding himself of the mosquito's indispensable place in the ecosystem. Without the mosquito and its larvae, most of the animal world of the Everglades and Keys upward through the food chain would sicken and die for lack of nourishment from below. And yet the mangrove country does seem to overdo its hospitality for mosquitoes.

But then the mangrove swamp seems to overdo its hospitality for all its citizens.

Beginning in the Keys and running around the southern end of both coasts of Florida, vast mangrove tangles for centuries have built land through a steady encroachment on the submerged continental shelf. Bulldozers have ripped out the land-building mangroves on most of the east coast to make room for boat slips and hotel sand beaches (and the sand beaches, unprotected by mangrove roots, have been swept away by winter storms so that sea waters now lap at the foundations of many "beach resort" hotels.) But up the west coast stretches the world's largest mangrove jungle as a vast brooder and nursery for a stupendous Gulf fishery and a rookery for tens of thousands of tropical and semitropical birds.

Inland from the coastal strip, in Everglades National Park, grow exotic trees, such as strangler figs, gumbo limbos, mahoganies, and the poisonous manchineels—all blown in from the West Indies on hurricane winds. But along the salt-water coast the red mangrove flourishes alone in one of the few jungles that truly merits being called impenetrable.

Standing on its roots, the mangrove proliferates in a frenzy of propagation. Its elongated seeds put out roots even before falling from the tree so that they can lock onto the first exposed lump of mud they encounter and push farther into the Gulf.

Though it tolerates salt water, *Rhizophora mangle* or red mangrove prefers a less saline habitat, so its tangled roots trap fresh water leaking down the vast River of Grass from Lake Okeechobee through the Everglades and hold it in a stagnant brackish mix with salt Gulf waters.

Spread to the semitropical Gulf sun, mangroves lock solar energy into their leaves through the miracle of photosynthesis. Leaves fall off into the soup around the mangrove roots. There anerobic bacteria break them down and release the sun's energy as nutrient chemicals. Silt washed down from the Glades is also trapped in the tangle. A fertile land slowly builds under the swamp and pushes it farther to sea. Sunshine filters through the dense foliage to speed the decaying process. And the helpless young of shrimp, crabs, and dozens of game and commercial fish species in unthinkable trillions drift into the tangle to feed on the nourishing broth of the sluggish waters and to hide from their enemies in the underwater jungle of mangrove roots.

Every one of the brown and pink shrimp taken by American trawlers and Cuban poachers around the Dry Tortugas grew to near maturity in the mangroves that make up the Ten Thousand Islands jungle up the west coast of Florida to Cape Romano. The shrimp double in size weekly on the rich diet of swamp waters.

To most outlanders, the mangrove swamp is repellent. When I was a young reporter I had to travel with sheriffs' posses through mangrove swamps that were a nightmare of fiddler crabs scurrying from our flashlight beams, scuttling lizards, spoil banks rotten with a musky smell the deputies said meant rattlesnakes, and the choking ammoniac smell of droppings around pelican rookeries. And above all the mosquitoes.

Deep in the mangroves they swarmed so dense that we had to tie handkerchiefs about our faces to breathe. A swat against a sleeve crushed a dozen or a score of mosquitoes. My eyes were swollen almost shut for a day after escaping from the horror.

But on a return to the Ten Thousand Islands years later, I drifted through on a boat, silent to avoid disturbing the roseate spoonbills then looking for nesting sites. Even a few feet out in the stream from the mangrove jungle, the air was salt-fresh and mosquito-free. Bird song and squawk spoke of a vigorous wildlife in the jungle, and the young of a hundred species of sea creatures swarmed through the clear shallow waters under our keel.

The mangrove islands continue a short distance north of

Cape Romano, seventy-five miles upcoast from Florida's southern tip, but the bulldozers have arrived there and the wilderness is fast disappearing. The cape marks the effective northern boundary of the wild mangrove world. Flying over the submerged flats of Gullivan Bay and Cape Romano Shoals, I was thrilled and astonished at the swarming sea life. Not an acre of sea bottom that did not support at least one shark or a ray with a finspread wider than a man's armspread. Fish swarmed in schools so dense they hid the bottom. Every sandbar was clamorous with densely packed seabirds—pelicans, gulls, terns, herons.

I had read just the night before in an environmental expert's doomsday lament for the death of Florida that the manatee was long since extinct on the Gulf Coast. I am happy to report that I watched a specimen at least seven feet long idly plough through a bed of sea grass, devouring bushels of the green stuff and leaving a trail of muddy swirls where its flipper tail ruffled the mud bottom.

Leaking out from the rich soup around the mangrove roots, a steady stream of sea creatures near the bottom of the food chain nurtured those thousands of predators that made such pretty pictures under the Gulf's placid surface.

Farther north, racing speedboats made pretty patterns on the water, but the sea creatures have gone, driven away when dragline and bulldozer converted the sheltering mangrove swamp to a waterfront city with every lot on its own boat slip. From the air, the boat cities with their long skinny building sites separated by dead-end boat canals look like stiff-fingered claws raking the life from the Gulf bottom.

And yet there is no denying that the waterfront cities offer an affluent life so comfortable and so outdoorsy in an ersatz way that only a fanatic environmentalist would banish them. I accepted the hospitality of the good folk on Marcos Island and Naples and enjoyed enormously racing over the waters, sipping a fine champagne, and worrying little about the dead Gulf bottom under our keel where once a teeming sea nursery flourished.

At some time in recent years, however, it dawned on state officials that the Dry Tortugas fishery alone was a billion-dollar business entirely dependent on those mangrove tangles and muddy flats that were being devoured by draglines to make waterfront developments. The state set up a Bureau of Beaches and Shores and a Coastal Coordinating Council to bring further shoreside construction under control.

Cliff Ellis, head of the Division of Marine Resources, told me that Florida's coastline had been lengthened by 8000 miles in historic times through the building of artificial harbors and finger canals for pleasure boats. And the Gulf Coast, though not nearly so fretted by finger canals as the Atlantic Coast, desperately needs protection.

"No sun can penetrate the murky waters of the canals. The young of the sea creatures no longer find a home where the dragline has torn out the bottom. Look at Fort Lauderdale on the east coast. It's a lovely city for the boating fan with two hundred miles of boat canals inside the city limits. But where are the tremendous schools of redfish, sea trout, and sheepshead that used to swim off the coast? Gone, man, and they'll never be back because their nursery is gone.

"But on the Gulf Coast where dredging and filling hasn't gone too far, even heavy commercial fishing doesn't put a dent in the sea populations."

North of Cape Romano to the beaches beyond St. Petersburg the bulldozer has worked overtime. An official of the Bureau of Beaches and Shores said early developers working before his bureau was established to protect the shoreline, built so close to the edge of the Gulf that they had to bulkhead against winter storms. So the storm waves, instead of spending their energy harmlessly in a long dash up a gradually sloping beach, crashed into the bulkheads and scoured out the remaining beach on their turbulent "long withdrawing roar."

Now builders must get permits for shoreline bulkheads which must be set back so far that beaches are not endangered.

"On the Atlantic Coast, especially at Miami Beach, many resort hotels rise so high that by two o'clock in the afternoon they throw a shadow over sunbathing guests. Not that it makes much difference, however, for most of the beach has been washed away anyhow. At Miami Beach it will cost at least forty million to dredge the sand from where it's been washed offshore and return it above the waterline where it belongs.

"On the Gulf Coast we have already had to restore the beach at Treasure Island. With luck, enforcement of new standards will protect what's left elsewhere."

Even the most impeccably careful tinkering with the environment can bring dismaying changes.

Sanibel and Captiva islands, for instance, have long been famous for offering charmingly isolated beaches where storms dump more than four hundred varieties of shells. Farthest south of the true barrier islands that line much of the Gulf Coast, the twelve-by-two-mile Sanibel was built by gradual deposit of shells ground by wave action to a coarse sand. Since the late nineteenth century shell enthusiasts have made the three-mile crossing from the mainland for the pleasure of solitary rambles on the beach.

Only true enthusiasts could stand the plague of mosquitoes

that infested Sanibel. Mosquitoes lie near the bottom of the pyramid of life everywhere on the Gulf Coast. But on Sanibel they truly found a home. Salt-water varieties laid their eggs on the tidal mud flats on the north side of the island; larvae survived the daily flushing and drying out of tidal surges, but predatory fish that normally hold down mosquito numbers could not survive the tidal ebb and flow. So mosquitoes proliferated out of all control.

During one night in 1950 on Sanibel a single mosquito trap caught 366,000 of the pests. Scientists estimated that breeding grounds carried 45,000 eggs per square foot.

Then a state board of health official attacked the problem. Using methods unassailable by the most finicky environmentalist, he eschewed pesticides that might threaten other forms of life. Instead, he installed canals and water gates to hold water on the tidal flats so that topminnows and killifish could get at the mosquito larvae around the clock. Now the mosquitoes of Sanibel zing about in sharply reduced numbers.

Not so the human visitors, however, for removal of the mosquito pest—plus construction of a causeway linking the island to the mainland—opened floodgates to a tide of solitude seekers. By the hundreds they pour across the causeway onto the famous shell beach where they circle about one another, searching for the precious solitude that has fled the island forever.

And the swallows that used to gorge themselves on mosquitoes at Sanibel during their migrations have developed insane flight patterns, possibly as a convulsive response to the disappearance of their wonted banquet. Jinxing, diving, zooming in a mindless frenzy, they dash themselves against automobile windshields in a suicidal fury.

The barrier islands and peninsulas from Sanibel and Captiva, separated from the mainland by narrow salt lagoons, support a booming resort development. (More sober businesses stick to the mainland shore.) From the tip of Pass-a-Grill Beach southeast of St. Petersburg through Clearwater Beach twenty-two miles to the north, motels, restaurants, and night spots stand shoulder to shoulder. An adroit broad jumper could almost travel the twenty-two miles of barrier beach by walking across rooftops.

Virtually all the resort structures are new and well designed, and the beach is clean and bright under the Gulf sun. But the great flocks of birds that haunt Sanibel's wildlife refuge—the herons and egrets, the roseate spoonbills and ibis—are nowhere in sight. Only tough brown pelicans and unsquelchable gulls still hang about.

The resort life looks comfortable—even inviting—but it has little to do with the beauties of the Gulf Shore. A sunlamp beside an indoor pool back home would give most of the visitors as much recreation at considerably less expense.

Beginning just north of that resort strip, however, the habitat defeats the developers—or has until now.

The sandy beach disappears, replaced by mud flats. Shallows extend for miles to sea. The continental shelf, a limestone base covered with a rich silt, slopes only about one foot in every mile. No tourist seeking a dip in the Gulf would stop along this coast, for he'd have to wade half a morning through knee-deep mud to find water to his shoulders.

But sea grasses thrive in that shallow bay snuggled under the Big Bend of Florida's Gulf Coast. And hiding in the sea grasses are trillions of marine creatures growing and putting on strength before venturing among enemies in open Gulf waters.

For some reason, hurricanes rarely sweep these waters. Even Hurricane Agnes which struck in June 1972 did only minor damage—for a hurricane, and saved its real fury for the northeastern states where the storm dumped three months of normal rainfall in 24 hours, making a disaster area out of the most extensive flood zone in United States history.

Tidal ebb and flow are slack. Wave action is negligible. Marine scientists call it a zero energy beach, and for that reason big resort developers have found it a zero-money zone.

But sports fishing camps proliferate, for the great sea-grass nurseries yield stupendous populations of game fish. Commercial fishermen have worked the Big Bend schools for decades without affecting the yield. Sports fishermen, even in much greater numbers than at present, could not make the faintest impression on the schools. Not so long as the mud flats lie undisturbed as gigantic incubators for sea life.

And the future looks good for the mud flats with several new state watchdog agencies standing guard.

Even dredging of a channel across the flats to extend the Intracoastal Waterway from San Marcos to Yankeetown would not doom the sea-grass beds. Conservation officials have plans to use the dredge spoil for building artificial barrier islands, shutting off an inshore lagoon from what little wave action now ruffles it and creating new grass beds on the seaward side. In effect, with careful handling to allow current flow and water exchange between lagoon and Gulf, dredging would create two zero-energy nurseries where one now exists.

It's along this same low-energy zone that the Gulf Coast begins to take on a different look. Mangroves disappear; pines creep down almost to the shore. Except for scrubby palmettos and the ubiquitous Sabal palm, the state tree of Florida, the vegetation takes on a continental look.

Just inland, swamps cover vast areas. Thousands of acres around the Big Bend have never felt the tread of man except possibly for some prehistoric Indian wanderer. Unlike most of the Everglades in the south, these Big Bend swamps support forests, and most of the land belongs to timber and pulpwood companies. The Tate's Hell region just inland from Carrabelle supposedly got its name because a certain Tate was lost in its fastnesses and turned whiteheaded before finding his way out. The Buckeye Cellulose Corporation owns the tract, and so it will inevitably be logged, but it still presents a formidable challenge to explorers.

The first serious effort at land exploration of what is now North America came to grief in those swamps. In 1528 a Spanish gentleman-gangster named Pánfilo de Narváez who had landed months earlier near Tampa Bay fled with his band of conquistadors to the coast near present-day St. Marks on Apalachee Bay. Using the swamps as a refuge and base for guerrilla attacks, Indians had harassed the flanks and rear of the Spanish column. The conquistadors might have stayed to fight if the Indians had been worthy opponents, but the savages had turned out to be miserably poor, an inexcusable defect in the conquistador code, and so the Spaniards were bugging out.

Normally scornful of any man so low of caste as to be a skillful workman, the hidalgos discovered in themselves an unsuspected mechanical genius when escape by sea became the only way to save a whole skin. Castilian and Andalusian noblemen who had never wrapped hands around any tool but a broadsword learned overnight how to fell trees, how to make saws from armor, and

◄ Little Mullet Key, an uninhabited island several miles west of Key West

Marsh mallow near the town of Everglades

Sunrise over a misty marsh near Everglades

Alligator, Cape Sable

Newly hatched brown pelicans in a mangrove thicket near the Ten Thousand Islands

Deer Point Lake near Panama City ►

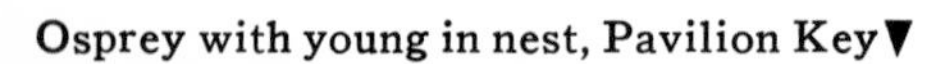

Osprey with young in nest, Pavilion Key ▼

▲ White pelicans, Sarasota

▼ Lilies, Apalachicola

◀ Flock of black skimmers in the small fishing village of Cedar Key south of the Suwannee River

Dunes east

rt Walton Beach

Shells, Pine Island Sound

Shoreline of an uninhabited island in Boca Grande Pass

Royal terns on one of Florida's few remaining undisturbed islands at the west end of Tampa Bay

Early morning in a bayou leading into Mobile Bay ►

how to use them to cut planks for boat sheathing. They made a crude forge and bellows to turn spurs and spearpoints into boat hardware. They made sails from their shirts. Their extremity even led them to the ultimate horror, for a caballero, of slaughtering their beloved barbs to make water bottles of the hides. They put out in four wobbly ill-made craft and turned westward, hoping to reach the new Spanish settlements of Mexico by crossing the Gulf. Only four of them, led by the intrepid and oddly appealing Cabeza de Vaca, the king's treasurer, ever reached civilization. The others drowned in storms or starved as slaves of coastal Indians.

Just around the headland of Alligator Point that closes in Apalachee Bay, coastal barrier islands and sand beaches predominate again, replacing the mud-flat coast of the Big Bend. At the Apalachicola River an extraordinary split in ecosystems occurs for no reason explainable by the present level of science. Creatures that flourish on the east bank do not exist on the west bank. Most of them are so inconspicuous that the casual traveler never notices the change. Who but a professional naturalist would remark the absence on the west bank of a two-legged salamander that inhabits the east bank? But the abrupt change in fauna and flora is no less amazing for being inconspicuous.

A striking change from the South Florida scene in recent years is the continuing health in the Panhandle of Spanish moss, that lends a graceful beauty to the most commonplace scenes. Waving from tree limbs and telephone wires, the harmless air plant softens the harshest scene so that even a tarpaper shanty looks romantic at the bottom of a moss-draped live-oak alley. Farther south, especially in the Ocala region, a blight has struck the Spanish moss. It has sickened and died so that the Ocala horse-farm country looks more than ever like the northern bluegrass fields of Kentucky. The horse farms with their neatly fenced pastures and gamboling colts can stand the passing of the moss, but much of South Florida's countryside sadly misses the softening effect of the blighted plant.

Spanish moss still drapes the Panhandle forests, which, indeed, means virtually the entire Panhandle, for woodlands cover the westward-pointing arm of the state as an almost solid blanket, pocked only rarely, away from the coast, by small villages where pecan culture and quail shooting are the prime topics of conversation around the general stores.

Along the coastal strip, however, the wind has sculptured sand into as pretty a dune landscape as the planet affords. From Dog Island off Carrabelle just past the Big Bend to the islands off the Mississippi's delta, a procession of majestic white sand hillocks marches parallel to the sea beside a hard-packed white sand beach. Named the Miracle Strip by local promoters, Florida's Panhandle coast has largely gone the way of the beach at St. Petersburg with an almost unbroken strip of beach resort motels. Here and there Air Force installations and other government holdings open the seascape to unvexed view.

Most notably on the Federal Wildlife Refuge at St. Vincent Island can the visitor walk a beach just about as wild as it was when the Indians came down from the interior for annual frolics as they smoked a supply of Gulf fish for the coming winter.

Touring the piney woods and cypress bogs of the refuge island's interior with Ranger Sims Vickery, I got my first attack of literal buck fever in twenty-five years of photojournalism. We rounded a bend in the forest trail and I froze. Beside the path stood a buck as tall as a young moose with a spread of antlers like a Colorado wapiti. With shaking fingers I fumbled with my camera till the great beast plunged off into the brush. Expecting only a gentle whitetail deer, I was dumfounded.

"Sambar buck from India," Sims said. "The folks who used to own the island planted all kinds of exotics here about 1905. Most of them couldn't make it in this strange country, but the sambars did fine. We probably got more sambars on the island than whitetails now."

Among the exotics once living on the island were African elands, noble antelopes of heroic size now being considered for domestication as a meat animal. The last eland known on the island swam to the mainland somehow and was very naturally slaughtered by the first hunter who got him in rifle range, though no season on elands has ever been proclaimed anywhere on the American continent.

"Stay alive and you might get another shot at a sambar picture," Sims said. "They haven't gone near so wild as the cattle the owners left behind when they sold the place to the government."

The huge inlet of Escambia Bay at Pensacola suffers repeated fish kills with a consequent flurry of accusations and recriminations, but nobody is really certain what is killing off the water creatures. Not every expert is even sure the fish kills are a strictly modern phenomenon.

But the kills have contributed one good to the area's future by helping to force development of a land-use plan to freeze further uncontrolled exploitation and to zone the coast and its inland waters for minimum damage to the environment and maximum preservation of remaining wilderness. Known as the Escarosa Plan, the zoning program may pioneer environmental protection for the whole state—eventually for the whole Gulf Coast.

The Bayou Country

Just east of the Florida–Alabama line begins the world's longest stretch of estuarine bays, lagoons, marshes, barrier islands, embayments, peninsulas, and river deltas, not excepting the Amazon's many mouths. Possibly the pullulating marine life of those shallow basins of brackish waters, guaranteeing a full larder, persuaded the first permanent settlers of the Gulf Coast in what is now the United States to establish townsites in Pensacola, Mobile, Ocean Springs, and Biloxi.

Apalachicola at the western end of Florida's Big Bend has superb oyster reefs—and indeed early Spanish settlers of Florida's Gulf Coast did build their cabins not far from there at San Marcos, facing the fecund *zero-energy zone* sheltered by the Big Bend's curving shores. But the ideal combination of deepwater ports and shallow seafood nurseries begins farther west at Pensacola and Mobile, most especially at Mobile Bay.

Mariners began seeking shelter and fresh water in Mobile

Bay possibly as early as 1519 when Alonzo Alvarez de Pineda, explorer of the Gulf's north shore, may have anchored there—or possibly it was in Pensacola's splendid harbor.

In 1528 the hapless Pánfilo de Narváez and his party, fleeing the ferocious Apalachee Indians in their makeshift flotilla, were seduced by the inviting gestures of Indians at Mobile Bay into coming ashore for a feast. Halfway through the festivities, the Indians pounced on their guests—a treachery that was, in the French statesman's phrase, worse than a crime; it was a blunder. For they had not reckoned on the terrible fighting skills of those battle-hardened conquistadors, veterans of a hundred savage battles in Europe and Mexico. The Spaniards carved their way back to their boats, leaving behind a badly mauled band of savages.

A generation later the young bloods of a neighboring clan tried their tomahawks on Hernando de Soto, with worse results, for even the hard-bitten Pánfilo de Narváez was a sissy compared to De Soto. The second band of conquistadors not only carved the Indians but burned the village after looting the granary and carrying off anything of value that wasn't pegged down—but not before losing seventy killed, and many of their stores in the fire.

(Serious historians can discount a still earlier visit—the twelfth-century arrival of the legendary Prince Madoc of Wales and a party of Welsh colonists—which landing, a sign erected by the Daughters of the American Revolution, announces as having taken place at present Fort Morgan, on a sandspit at the east side of Mobile Bay. Nobody had ever heard of Prince Madoc's American explorations till English historians conveniently discovered them to deny Spanish first claim to the vast New World based on Columbus's voyages.)

Alabama's stubby share of the Gulf Shore is only about fifty miles across, but the deep indentation of Mobile Bay gives it 397,353 acres of valuable estuary plus 34,614 acres of tidal marsh and 433 miles of bay and open-water shoreline—ideal country for a sea-life nursery. The city of Mobile—a strikingly beautiful city that looks like a movie set for an old Southern town—enjoys a bustling industry and scarcely knows of Mobile Bay except as a port to handle the city's commerce. But outside of the city and its suburbs most residents of Mobile and Baldwin counties around the bay—perhaps a hefty 70 per cent—make their living from the sea as boatbuilders, fishermen, seafood processors and shippers.

Almost immediately across the Florida line in Alabama, the village of Bon Secours on Mobile Bay markets as succulent an oyster as any American shore affords. In the moss-hung streets of the town you can spot the oyster fishermen because years of swinging the cumbersome stales of their oyster tongs have broadened their backs so that they look like those of heavyweight wrestlers. Operated like posthole diggers, the tongs make excruciating labor of the oyster harvest, but conservation laws forbid mechanical dredging except on about five hundred acres leased from the state and farmed as intensively as a tobacco patch.

Across the bay at Bayou La Batre, a fleet of trawlers and a boatbuilding yard make a handsome dollar out of the shrimp that grow to maturity in the shelter of the estuary and then venture to the deep waters of the Gulf. An 85-foot steel trawler costs $150,000, but an industrious owner-captain can pay off the loan in five years. Indeed, local banks look askance at any fisherman asking for longer terms. An ordinary deckhand makes $1100 a month during the season. And since his main job is sorting the catch to throw trash fish over the side, the only skill or education he needs is the ability to tell a shrimp from a crab, a flounder from a sand shark.

(Presently, shrimpers throw away 90 per cent of their catch, a shockingly wasteful process doomed by economics. Though Alabama marine laboratory scientists say no species is being overexploited, and the total catch increases a steady 15 per cent annually, experience on the Grand Bank fisheries and elsewhere shows that strained marine resources are forcing less selective fishing; many species once scorned will soon be harvested. Shrimpers now throw away all skates, for instance, but skates are highly prized in European kitchens and many an expensive scallop dinner in the United States started life as an humble skate.)

There is a catch to the Alabama fisherman's apparently paradisaical lot. More than 85 per cent of the shrimp catch comes to market from June through October. During the other seven months of the year the shrimper pumps gasoline or does odd jobs for cigarette money.

With the discovery by the Bureau of Commercial Fisheries of a large population of swordfish in the Gulf, one Alabama shrimper converted his 80-foot boat with the hope of having a year-round job. His first cruise brought him an encouraging 14,000 pounds. But the Food and Drug Administration condemned the lot for excessive mercury content. So the shrimpers have resigned themselves to sweating for five months and loafing about for seven while the swordfish lurk outside uncaught.

Halfway down the bay road from Mobile to the Gulf, Bellingrath Gardens offers 65 acres of exquisitely manicured formal garden in year-round bloom for tourists. More important to the wildlife, the gardens also offer 725 acres of undeveloped bayou and marsh country for sheltering migratory birds. With the broad Gulf crossing ahead of them in the spring, birds stay over in the gardens to fill up on berries, cracked corn, and the native insects that spawn in unimaginable billions in the Alabama estuary. In the autumn birds tumble into the gardens exhausted and famished after their long overwater flight.

Logbooks record more than two hundred bird species—including a pair of swans imported from Europe and a muster of flamingos from the Caribbean. The swans and flamingos have found a comfortable home. To assure the flamingos' food supply, Boy Scouts released in garden ponds several hundred pounds of crawfish caught in surrounding marshes. The exotics try to nest and establish themselves as part of the Alabama scene, but foxes and raccoons make a good thing out of the newcomers' ignorance of Alabama wildlife ways. The rare nestling that hatches doesn't survive long enough to feather out properly.

Of all the barrier islands along the Gulf Shore, none has more beautiful dunes than Alabama's Dauphin Island. I cannot understand the loathing for the place expressed by Antoine de la Mothe Cadillac, the second governor of French Louisiana, who called it in the first years of the eighteenth century ". . . sand, fit only for hourglasses." The islanders still resent his unkind appraisal of their dunes homeland. Two and a half centuries later a mural at the Alabama marine laboratory on Dauphin Island publicly denounces Cadillac's administration and Indian policies, and a sign marking the former site of his residence sniffily makes note of his low esteem for this sandy capital.

Something about the Mobile Bay country made the French settlers restless. In 1699 they had landed near Biloxi to the west in what is now Mississippi and had set up a capital to govern the vast French Louisiana territory that took in the lands between the Appalachian and Rocky mountains and from the Gulf to Canada. Almost immediately they moved to Mobile, but in less than twenty years they unaccountably moved again, winding up in New Orleans on a boggy site that offered not nearly the access to the sea or the comfortable anchorages of Mobile Bay.

Westward from Alabama, along the seventy-mile stretch of Mississippi Gulf Coast, for years to come the scene will be dominated by the catastrophe of hurricane Camille. In August 1969 Camille blew out of the Gulf and tore the Mississippi coast to tatters.

The highest winds ever recorded in a hurricane in North America blew away the coastal highway, hundreds of thousands of venerable live oaks and pines, hotels and motels, supermarkets and restaurants, quays and piers. Trawlers rode the sea surge inland and were stranded. Nobody knows how many died, for some of the victims were of those humble folk who are almost faceless, but the number certainly ran more than two hundred. A party of friends who gathered in a seafront apartment building to defy the storm with a hurricane party drowned to the last man when the building was demolished.

Years after the blow, streets have been repaved and debris long since cleared away. Motels, restaurants, and night spots have been rebuilt. But along the waterfront, once graced by aging Southern mansions, now stretch weedy gaps empty except for an occasional concrete flight of steps showing where somebody's home once stood.

The very sands of the beach have never recovered their former dazzling white. But the beach can be restored, for it was man-made to begin with and can be man-made again.

For generations, planters from the sun-baked interior came to the Mississippi Gulf Coast for cooling sea breezes, but the mediocre beaches were never an important lure. Then the Corps of Engineers in 1951 dredged an offshore channel and dumped the spoil on the resort's front doorstep. Innkeepers set up an angry howl as the sickening black sludge covered what little beauty their beach had once had.

But within weeks a seeming miracle happened. Whatever the particles were that gave the spoil its revolting texture and color, they leached out with tide and rain, leaving behind a twenty-eight-mile beach of sand as white and fine as confectioners' sugar.

Mississippi has made one important gain on its coast since Camille. The federal government took in the offshore barrier islands as a National Seashore to protect the relatively undeveloped sandy keys against uncontrolled exploitation. Visitors can go by boat to Petit Bois, Horn, and Ship islands to see how the whole coast looked before the Sieur d'Iberville brought the first load of settlers.

The seashore park would include one more island except for the overeager attentions of a Chicago wholesale florist. He offered a handsome price for mature heads of sea oats, and so housewives from the mainland flocked to Caprice Island with scissors and baskets to reap the golden harvest. Too late, inhabitants of the coast, including the owner of an elaborate resort hotel on Caprice Island, discovered that sea-oat roots are all that hold the barrier islands and the mainland dunes together. Stripped of plant cover, the Isle of Caprice blew away on every vagrant breeze and the hotel fell into the sea. All that's left now is an upwelling bubble that ripples the Gulf's surface where the hotel's artesian well still spurts a stream of fresh water into the sea.

(Warned by the fate of Mississippi's island, Florida has posted signs along its Gulf Coast warning that picking of sea oats is against the law.)

Just over the invisible state line in Mississippi, the offshore islands are also melting away, victims of technology. To hold the capricious Mississippi River within a predictable course, the Corps of Engineers has locked it inside levees. By a quirk of geology, the present river mouths pour their waters into the sea at the closest approach to the edge of the continental shelf of any coast in the Gulf of Mexico. So the silt stripped from thirty-one states and three Canadian provinces no longer builds land, as it did for 5000 years, but dumps it into the abyss. Every high sea erodes a bit more of the Louisiana coast. With the river's silt no longer replacing eroded land, the state shrinks 16.5 square miles every year.

Flying with Max Summers, assistant chief of Louisiana's oyster protection division, I could see striking evidence of land loss. On the map, the Chandeleur Islands form a strong barrier crescent protecting the fragile marsh of the bird-foot delta. In fact, from the air they are barely visible and for much of their length show only as ruffled whitecaps marking a now sunken sandbar.

They still shelter and nurture fish, however, for we jumped a pair of Alabama skiff fisherman illegally putting out nets in Louisiana waters. Max summoned enforcers by radio and they scrambled in an amphibious Widgeon airplane to catch the poachers while they were still at work.

"These shallows near the delta are the main oyster beds of the continent," Max said. "We cannot afford to let them wash away."

We flew over Bayou Lamoque where a siphon scoops up a trickle of river flow, lifts it over the levee, and dumps it into Breton Sound between the mainland and the shrinking offshore islands. A brown fan of silt spread from the mouth of the bayou, staining the clear waters of Breton Sound with life-giving nutrients carried downriver by the Mississippi from the heart of the continent. Six more siphons to pour silt into the marsh will soon try to rebuild lost marshlands and preserve the richest seafood nursery in the world.

The siphons represent as clear a case as I've ever encountered of a head-on collision between well-meaning agencies, with the Corps of Engineers struggling to keep the river between its banks and the state conservation agencies siphoning the river across its banks in artificial flood.

"Louisiana has five million acres of marsh and twenty million acres of offshore shallows, all of it the gift of the river," Max said. "They are the richest lands on earth. A salt marsh produces more plant material than a well-fertilized wheat field. Without any farming whatever, an acre of salt marsh produces ten tons of dried straw and the foodstuff to nourish by far the most important fur and fish industry in America."

I protested that Louisiana fisheries could not exceed Alaska where I have witnessed torrents of salmon pouring into canneries working around the clock to handle the immense harvest.

"Maybe not in dollars but certainly in pounds. About one-third of the whole American catch comes from the Gulf," Max said, "and Louisiana takes fifty-four per cent of that, more than all the other Gulf states put together. Our fishermen every year bring in twenty million pounds of edible fish, twelve million pounds of oyster meat, ten million crabs, eighty million pounds of shrimp, about a billion pounds of menhaden, and twenty million pounds of rough fish to feed ranch mink....

"And trappers bring in up to ten million pelts a year, mostly muskrat and nutria, with a few coons, minks, and otters thrown in.

Their value was almost forty per cent of the American catch, far more than any other state, including Alaska."

We landed at Lafitte's old hangout on Grand Terre to visit the state's marine biology laboratory for a talk with Dr. Lyle St. Amant, one of the world's foremost experts on conservation of sea life. Pointing to the offshore platforms of oil and sulphur companies, clearly visible from the beach, I asked how much damage mineral exploitation had done.

"Offshore, virtually none, that I know of. In fact, seventy charter boat captains do a pretty brisk business now fishing around the rigs where we didn't have a single boat a few years ago. The rigs provide a kind of artificial reef habitat. Sea plants and barnacles start growing on the piers, the little creatures that live off them come around, and naturally the bigger things that eat the little things follow. Soon you have a flourishing colony where it used to be just open empty sea.

"Cutting canals and throwing up spoil banks at well sites ashore have done some damage. But oil exploitation started here almost fifty years ago. We have more than twenty-four thousand active wells between the three-mile offshore limit and the north edge of the marsh, and another fourteen thousand wells beyond the three-mile limit. And still we lead the whole nation in fish and fur production.

"The two industries of oil production and fisheries are apparently able to coexist and maintain near-maximum production under present regulations."

From the laboratory windows we could see a sandbar in the bay where imported brown pelicans had tried to nest. Partly because the brown pelican is Louisiana's state bird, extinction of the species on that shore has embarrassed Louisianans. So the game department brought in 175 Florida birds and released them under the watchful eyes of the marine scientists.

Though they all came from tree rookeries in Florida, they built their nests beside the water on a shallow sandbar. The first wind-driven high tide floated eggs and nests away. So only seven birds have hatched and grown to flying size in the transplanted colony, but the big birds were rebuilding nests and still trying the last time I saw them.

West of Grand Isle we flew over a vast empty swamp, the Atchafalaya Basin along the banks of the Atchafalaya River. In the 1950s the Atchafalaya threatened to capture the Mississippi River at a point near Natchez, Mississippi, and take it to the Gulf by its 175-mile shortcut route. Repeatedly the Mississippi has shortened itself after building an awkwardly long delta.

The present course is only about six hundred years old. But since Baton Rouge and New Orleans grew up on the lower riverbanks, the country can no longer afford to indulge the river's caprices. So engineers closed off the link between Atchafalaya and Mississippi and allow only about one-third of the upriver Mississippi water to escape by the shortcut. That one-third is enough to carry a tremendous load of silt to the Gulf, however, and an important geologic event is happening out there. From the plane I could clearly see the vast load of mud settling in Atchafalaya Bay, building a new bird-foot delta and restoring some of the marsh and estuarine nurseries being stolen by the sea elsewhere. By 2020, according to geologists, the entire bay will become land—if you can call a jellylike mass held together by roots of marsh grasses so solid a name as "land."

Upstream the same silt has conservationists and engineers in an uproar. Inevitably it falls out of the stream during periodic overflows and the basin is silting rapidly at the northern end. Hunters, fishermen, bird watchers, conservationists in general watch in anguish as their beloved wild swamp disappears. An inconclusive battle rages over the best way to stop loss of the wetlands. While tempers flare, the silt inexorably drifts down the river and settles on the bottom to make soybean fields out of bass holes.

West of the Atchafalaya stretch unbroken miles of salt marsh, most of it in public and private wildlife refuges. In virtually every pond I saw at least one giant alligator basking in the sunshine.

The comeback of the alligator from near extinction in these refuges makes one of the few encouraging stories in the environmental crisis. Al Ensminger, Louisiana's chief of refuges, told me the Rockefeller Refuge filled to alligator-carrying capacity with only ten years of protection.

"In the western parishes we could use a limited harvest, controlled by a license and tag system like that of the fur-trapping industry.

"Alligators are a renewable resource—we've proved that—and you have to face some hard facts. If you don't allow a landowner to make a dollar from the alligators his land supports, he's going to consider them a pest that eats up his profits in nutria, muskrat, and coons. Then he's going to get rid of them one way or another.

"We've got the alligator situation under control. It's time to let up on the reins a bit."

◄ Along an estuary of the Mobile River

Bayou La Batre, Alabama

Live oaks in an Alabama marsh

Ferns unfolding in a marsh bordering the mouth of the Texsaw River

◄ Beach near the Florida border

Bayou meandering through the marsh near Kreole, Mississippi

Flamingos, Bellingrath Gardens, Alabama

Wild iris, Cocoderie, Louisiana ►

Low tide on the beach near Gautier, Mississippi

Horn Island, part of the Mississippi National Seashore

Ghost crab, Horn Island

Shadows on the Teche, New Iberia, Louisiana, a property of the National Trust for Historic Preservation ►

Mushrooms, Ocean Springs, Mississippi
Panus Rudis, an edible fungus, Ocean Springs

Marsh grasses, Bay St. Louis

◄Gulls and terns at dusk near Biloxi, Mississippi

Horn Island, Mississippi National Seashore

Pitcher plant

South Pass, mouth of the Mississippi River

Coastal marsh near Chauvin, Louisiana

Chandeleur Islands

Spider lily, Rockefeller Wildlife Refuge east of Cameron, Louisiana

Freshwater crawfish

Cypress knees, Atchafalaya Basin

Canada goose, marsh near Grand Chenier, Louisiana

◄ Geese gather every morning to fill their craws with sand from the only sand bar in miles at Hell Hole, on the north shore of Marsh Island, Louisiana

Eggs of a wading shorebird near Grand Isle, Louisiana

Louisiana coastal marsh

Grand Chenier, near Cameron, Louisiana ►

The Texas Shore

For years I have driven highways paralleling the Texas coast and have come away with a blurred impression of sun-baked main streets lined with beer and bologna quick-service groceries and dusty pepper trees. I shall be forever grateful that I finally flew the length of the coast at low level with an amiable pilot who circled every flock of geese and gam of porpoises that caught my eye. Imprisoned between roadside ditches, the motorist can have no idea of the natural beauties of Texas lying just out of sight beyond the billboards.

Immediately after take-off near the Louisiana line, we began to spot the debris of hurricanes that have lashed that coast. All the way to Mexico the beach is littered with the rotting hulks of shrimp boats and the rusting cadavers of automobiles. For no good aesthetic reason, the wrecked boats are picturesque and the car bodies hideous. But mercifully the cars sink into the sand and crumble to red dust in the salt sea breezes.

South of Freeport the Brazos River stained the Gulf brown with a torrent of mud. Mixed with the load of silt there must have been a rich freight of nutrients, for white pelicans flew rigid formations like navy ships-of-the-line, searching for schools of fish feeding on river debris.

Flocks of waterfowl increased steadily in size as we flew southwestward. On the marshes around San Antonio Bay vast gaggles of snow geese mingled with dazzling flocks of roseate spoonbills. Only weeks before, the geese had been neighbors of walruses and seals resting on Arctic Sea ice.

The pilot made a wide detour to avoid flying over Aransas Wildlife Refuge so as not to disturb the fifty-nine whooping cranes wintering there. He dropped almost to the deck to skim over the site of the old city of Indianola, once the major port on the Texas coast, now gone, blown away by successive hurricanes. Like Omar Khayyám's wild ass, a bevy of spotted fallow deer now stamps over the heads of the vanished Indianolans, but cannot break their sleep.

The deer, imported from Europe, are only one of scores of exotic herds wandering in Texas. Even on the Aransas Wildlife Refuge a few fallow deer survive from a herd transplanted there by a former owner. The European wild boars released at the same time promptly bred with feral domestic swine, however, and produced a now stable subspecies. California quail and pheasants did not find a niche in the Texas Gulf ecosystem.

The migratory waterfowl found a home, though, for on the lagoon between Matagorda Island and the mainland and in the waters around St. Joseph Island vast rafts of waterfowl floated between oil-well platforms. Farther south every spoil bank thrown up by dredges digging out the Intracoastal Waterway had been converted by waterfowl to a teeming island rookery.

At Padre Island we found a lone whitetail deer and a single

coyote loping apparently aimlessly across a barren sand flat rippled by the wind like a washboard. On the beach we passed several campers up to the hub caps in sand, their drivers waving signals at us to send help.

Once again on the ground, I headed back for that same coast with high expectations, knowing that by leaving the imprisoning highway at selected areas I would see a Texas I hadn't suspected.

Beyond the Sabine River boundary in Texas my car was stopped on the coastal highway by a mass of lowing cattle, driven by half a dozen cowhands on pinto ponies. A pall of industrial smog from the Beaumont–Port Arthur complex smudged the sky, but the corniest Hollywood director could not have set up a more home-on-the-range scene than the ambling cattle and the denim-clad herdsmen wearing ten-gallon hats and dangling cigarettes from their chapped lips. Languidly lifting a single finger in salute and thanks for my patience, the cowboys passed, apparently unworried by the storm clouds gathering in the west.

A few miles down the coast the storm broke. Sleet peppered the windshield, and the cloud rack scudded low across the road, blocking view of the Gulf only a few hundred yards to the east. Disturbed by a thrilling call half-heard above the flapping of the windshield wiper, I stopped and got out to listen.

Floating on the wind came the mutter and honk of wild geese, a sound that speaks to something primordial surviving deep in all men. The skin-clad ancestor hidden within us listened to that same call and fitted an arrow to his bowstring in hope of bringing home a plump bird for his mate.

The ragged skein broke through the clouds and settled earthward in a spiraling vortex. Two thousand or more geese landed in the marsh not a hundred yards from the highway. Once landed, they disappeared, so that a motorist who had missed their arrival would pass the spot without knowing that he was within an arrow's flight of a vast flock of wild waterfowl.

Alerted by that gaggle of mixed snows and Canadas, I searched the marsh as I drove and discovered that the coast, apparently lifeless at first glance, was teeming with wildfowl feeding on marsh grasses. Even when they were hidden in the foliage, by stopping and listening intently I could hear the peculiar humming that signals a nearby flock searching the marsh for fallen seeds. Virtually every rice field along the coast had its wild gleaners. In a day's drive during the winter season any bird watcher can spot twenty thousand or more geese along the Texas coast.

Which makes all the more puzzling the near escape from starvation of Cabeza de Vaca during his years as a captive on this coast. After fleeing the Indians of Florida in Pánfilo de Narváez's rickety navy, in 1528 Cabeza de Vaca was cast ashore in a storm and enslaved by the Indians of Texas. He was lucky to fall into the hands of a comparatively gentle people rather than the cruel Karankawas of a few miles farther south, but his captors had to be incredibly primitive even by standards of the day and place.

Surely the game population of the sixteenth century was at least as numerous as today. And yet the Indians were resigned to losing a certain percentage of the tribe to starvation every winter.

His masters set Cabeza de Vaca to digging up with bare hands edible tubers on the roots of water plants. He spent winter nights nursing hands torn and bleeding, half-frozen from daylong immersion in mud and water. So famished were the Indians by the end of winter that Cabeza de Vaca escaped them by simply walking away while his masters were gorging themselves insensible on prickly pear fruit.

And all about them swarmed those wintering flocks of ducks and geese, fat and savory even for a well-fed man and the very stuff of life for a starving Indian.

Commercial fishermen today take two hundred pounds of seafood per acre from the shallow bottom of Galveston Bay, where Cabeza de Vaca probably starved through six winters. Tens of thousands of tons of oysters as succulent as any in North America grow there. Redfish and sea trout crowd the presently polluted waters, and so I can only guess at the vast schools that must have swam there four centuries ago. (Although at least two marine biologists have suggested to me the unorthodox theory that sea life may be flourishing on Galveston Bay's so-called pollution.)

The sixteenth-century Indians starving in the midst of plenty are not alone in being surrounded by unsuspected treasures around the Gulf Shore. Until recently nobody knew that a magnificent coral reef, now called the Flower Gardens, pushed within 50 feet of the Gulf surface 110 miles off Galveston's shore and hundreds of miles north of where coral should grow. Bob Alderdice, deputy director of the Flower Gardens Research Center, outlined for me plans to set up an underwater laboratory on the reef.

"Soundings on charts of the Gulf are almost pure fiction," Bob said. "Salt domes push up everywhere in the Gulf and for all we know corals may have established themselves on other domes besides the one lying under the Flower Garden reef.

"The gimmick that planted and nourished the Flower Gardens so far north of their normal habitat is the layering of water. From one hundred and fifty to two hundred and fifty meters deep flows a current of water from the Caribbean. When it strikes the fast-rising shelf just outboard of the Flower Gardens salt dome, it slides up toward the surface, carrying warm water and Caribbean-type nutrients. The coral creatures never know they have left their Caribbean home."

As a curious sidelight, Bob said that between 950 and 1050 meters down flows a river of Antarctic water.

The Marine Biomedical Institute at Galveston plans to build a laboratory over the Flower Gardens, similar to an offshore oil rig but with four underwater laboratories.

"A boat, used as a laboratory, is a lumbering dinosaur," Bob said. "On our boat yesterday at the Flower Gardens, during that same storm you drove through, all we could do was lie in our bunks and hang on. And a boat never really goes to the same place twice. It's impossible to get a true continuing sample. You're starting from scratch every day.

"But on our oil-rig type of platform we'll be on precisely the same station three hundred and sixty-five days a year and steady enough to use a microscope in any blow short of a hurricane."

Lost treasures of quicker currency than scientific knowledge obsess a sturdy percentage of the Texas coastal population. From Galveston, where four of the pirate Lafitte's ships went down in an 1818 hurricane, to Padre Island near the Mexican border, where a Spanish treasure fleet sank in 1553, beachcombers are forever finding rusted hidalgo spurs, bits of ancient chain mail, and even the occasional doubloon. Newspaper stories about the find set off another round of scuba-diving and beach-walking in a search for Spanish gold.

Treasure hunters rarely find anything of great value, but whether or not they find a sack of doubloons, on a beach walk they should see more bird species than anywhere else between Florida and California, for eastern and western, northern and southern habitats overlap, funneling in birds from the four points of the compass. Even the whooping crane chooses that strip of coast for its last stand on earth, though the marsh is marginal as crane habitat and inferior to the nearby Louisiana coast.

Most cosseted of all threatened species, the whoopers barely hang on to existence under the watchful eye of rangers and most of the coastal population around the Aransas Wildlife Refuge. Though fifty-four adults flew north the spring before I visited the refuge, they brought back only five juveniles, a poor return for all the care lavished on them.

Once common throughout the continent, even on the Pacific Coast, the whoopers began to disappear in 1871 when the last crane left Florida. Bird watchers reported the last nest in Illinois as early as 1880. From 1895 on, the American Ornithologists Union annual reports showed a steadily shrinking population and habitat. Largest of the colonies, the whoopers at Pine Island on the Louisiana Coast, disappeared utterly in 1900. Nesting failure reports moved steadily northward, from Illinois to Minnesota to North Dakota and across the border into Canada. It was only recently that the nesting grounds of these last survivors were found in Wood Buffalo National Park in Alberta.

Each year as the whoopers wing over the Dakotas, Nebraska, Kansas, Oklahoma, and Texas to the wintering grounds, conservationists log them in, praying that no sportsman with more magnum in his shotgun than woods lore in his head has shot down one of the magnificent birds.

One hopeful sign is that the whoopers seem to be loosening some of the rigid behavior patterns that make them susceptible to instant extinction by localized storms or other accidents.

Aboard the excursion boat *Whooping Crane*, Captain F. M. Brown, who has been carrying bird watchers to the wintering grounds for a decade, told me the whoopers are losing some of their old family habits. Till now each couple claimed about five hundred acres of marsh and stoutly defended their tract from intrusion by other adult whoopers. Recently, however, Captain Brown reported seeing birds freely violating neighboring territory and feeding up to their bellies in water as they had never done before.

During our trip Captain Brown pointed out triumphantly a group of six whoopers feeding together.

Elsewhere, I heard of crane sightings on Matagorda Island and even as far south as the Laguna Atascosa Wildlife Refuge near Brownsville.

The main concentration, nevertheless, remains on the Blackjack Peninsula of the Aransas refuge where, during the wintering season, rangers keep intruders out—including oil exploration teams that work while the cranes are north on the nesting grounds. But elsewhere in the refuge roam 3000 deer (down from a disastrous 12,000 population that was decimated by starvation and disease brought on by overbrowsing), about 300 hogs descended from a cross between imported European wild boars and feral domestic pigs, wild turkey introduced from woods across the Mississippi River, several drifts of javelinas, and even a bevy of fallow deer imported from England in 1931. The usual predators abound unmolested, though the red wolf has apparently bred itself out of existence by casual matings with coyotes and wild dogs.

Because cattle overgrazed the range before it became a refuge, live oaks grow in stands so dense that they never become more than runty bushes which game managers mow back to let competing grasses grow. Within a few years they hope to restore the original mixed oak-forest-grass savanna habitat.

While game animals wait for restoration of more open country, they use paths cut by oil exploration crews and clearings about gas-well sites to move through the stunted live-oak brush. In one small clearing about a Christmas tree wellhead, I counted four turkey gobblers and twelve whitetail does.

On a ten-mile drive along the Aransas Bay shoreline with Mrs. Doris McGuire, a veritable tiger among bird watchers, we spotted 2 oyster catchers, 5 killdeer, countless royal, Caspian, and Forster's terns, laughing, ringbill, and herring gulls, about 50 willets, 140 marble godwits, 125 black skimmers, 1 reddish egret, 2 great blue herons, 30 pintail ducks, a few American widgeons, 5 common golden-eye ducks, a kestrel, and uncountable myrtle warblers.

Curiously, in this congress of birds, Mrs. McGuire has never seen crows, blue jays, or starlings, though she has heard that thousands of jays once descended on the coast for a brief period and disappeared as mysteriously as they arrived.

At the Welder Wildlife Refuge, a private research operation a few miles inland, Dr. Clarence Cottam guided me about his domain.

"We've counted four hundred and fifty bird species and fifty-five mammals on the place," he said. "I've logged a hundred and forty-seven bird species in one morning."

In a bosky glade I counted seventeen tom turkeys on one side of a ridge and thirty hens on the other slope. Lurking just out of sight of the others, a lone tom searched among the fallen oak leaves for insects and mast.

"There's a sad fate," Dr. Cottam said. "Turkeys fight ferociously within flocks to establish a pecking order. When the chief is finally whipped by a young brave, all his old rivals remember past humiliations and drive him mercilessly out of their society to wander lonely as King Lear on the heath."

A flight of shoveler ducks flew off a pond and across the road. Literally like a bolt from the blue, a peregrine falcon swooped and knocked a female shoveler kicking and squawking into the marsh. Reflexively, I made to go to the duck's rescue, but Dr. Cottam stopped me.

"The duck belongs to the hawk," he said. "You can't enjoy the beauty of peregrines without allowing them a few ducks."

The role of the predator is one of the very problems being studied by the refuge research team, most of them graduate students working toward doctorates. Dr. Cottam was then fencing in a thousand-acre plot within which he would destroy all coyotes, bobcats, and other predators capable of pulling down a deer. Dynamics of deer population free of predation would then stop being a subject for argument between conflicting conservationist schools and would become a matter of proved scientific fact.

In that gentle climate even in winter I picked my way with caution through grassy stretches so as to avoid disturbing rattlesnakes. My wariness set Dr. Cottam off on a chain of snake stories.

"I once shot a fifty-five-inch cottonmouth mocassin and shook out of his stomach a fifty-nine-inch rattlesnake. What a battle that must have been! After I reported the event in a scientific paper, I got a note from a former secretary back East chiding me for having become a Texas tale stretcher so quickly.

"For the greatest battle I've ever seen I had an eminent witness to back me up. Guy Emerson, the head of the National

Audubon Society, was with me when we came across a hundred-and-two-inch indigo snake grappling with a fifty-five-inch rattler. There's been a lot of argument about an indigo's immunity to rattler venom. I saw the rattler sink his fangs deep into the indigo's side. The indigo swelled around the fang marks, but never slowed even slightly.

"The rattler knew only too well what was coming, and he tried to stave off dinnertime by stretching his jaws as wide as possible to make himself too big to swallow. So the indigo clamped his jaws just behind the rattler's head and inched forward, gradually forcing the mouth closed. All the while he pounded the rattler's head on the dry ground and squeezed him around the heart to make him more cooperative. He swallowed the rattler in minutes."

South from Corpus Christi to the Mexican line stretches the 120 miles of Padre Island. Nowhere more than three miles wide, the barrier island closes in the Laguna Madre, a shallow super-saline estuary that shelters astronomical numbers of waterfowl during the winter. A single raft of redhead ducks can stretch five miles. Every sandbar is covered with white pelicans and cormorants. Geese cover the marshes. Roseate spoonbills make brilliant splashes of color against the sky.

Driving from Corpus Christi to the north entrance of the Padre Island National Seashore, Barbara Shelton, park naturalist, took me through the Nueces County State Park to show me the impact of dune buggies on barrier islands.

Where the off-the-road vehicles race, snow fences have failed to hold the drifting sand; bulldozers must keep the paved road open; 700 feet of emergent shoreline have crowded into the Laguna Madre.

"The sea oat is the primary dune stabilizer," Mrs. Shelton said. "When dune buggies hit the foredunes, the oats go, the sand starts to blow, and before long the barrier island is headed for the mainland, stripping the mainland of its storm protection and filling in the valuable nursery grounds in the shallow lagoons."

Within the National Seashore wheeled vehicles must stay on the hard-packed beach, and sea oats in dense stands hold the dunes in place. (A few miles south of the main use area, an oil company land leveler stirred up a swath of sand eighteen miles long to obliterate truck tire ruts and return the beach to its original condition. Something in the process appealed to whatever seeds blew by and a superb crop of sea oats is currently building a new eighteen-mile-long dune.)

"Originally the island may have been wooded," Mrs. Shelton said. "In 1946 workmen uncovered a stump forest."

She pointed out a stand of five live-oak trees, the only trees in the entire park.

"They can't reproduce because the drifting sand buries the seedlings."

Cattle have grazed the land since about 1800 when the Portuguese padre who gave his title to the island brought in the first herd. Bluestem was then the best forage for cattle, but heavy overgrazing drove it from the scene. Marsh hay cord grass replaced the bluestem. But grazing was cut off in 1970, and almost immediately the bluestem began a swift recovery.

"Within a few years the original grasses will be dominant again," Mrs. Shelton said.

During my visit all park employees had been alerted that falcon hunters were working the area. Already a threatened species, the peregrine falcon suffers attrition from hunters armed with baited nets, for rich sheiks of the Middle East prize the birds and pay as much as $1500 for an outstanding specimen.

"Every year I get letters asking me where falcons congregate so the writer can net a specimen for scientific study," Mrs. Shelton said. "The only catch is the letters are invariably written in semiliterate English much more suitable for a hawknapper than a scholar."

Texas has 5000 species of wild flowers, more than any other state, and even the apparently inhospitable sands of Padre Island support six hundred species of plants. Since the disappearance of the live-oak groves, however, most of the animal life is nocturnal and furtive, like the blacktail jackrabbit or the ground squirrel. The lone coyote and whitetail deer I saw from the plane were rare specimens of mammalians larger than rodent size.

On the islands and spoil banks in the Laguna Madre between Padre and the mainland, however, swarm rookeries as densely populated by waterfowl as any other on the continent, including a rookery of the endangered reddish egret on the Audubon Society's South Bird Island.

The vulnerability of endangered species that insist on gathering in dense colonies was underlined a few weeks after my visit when a hailstorm with devilish precision pelted only the ten acres of the Audubon rookery, leaving the surrounding territory dry and undamaged. The hailstones smashed eggs, killed and crippled adult sea birds by the thousands. When the clouds blew over, dead birds littered the ground and surviving juveniles wandered aimlessly, peeping in their bewilderment.

Providentially, the storm hit while many more thousands of sea birds—herons, ibis, snowy egrets, and roseate spoonbills, in addition to the reddish egrets—had not returned from daylong feeding in mainland fields. So the reddish egret has suffered a setback, but so long as the habitat remains undamaged, most species can make near-miraculous comebacks from even worse disasters.

The eighty miles of coast in the National Seashore park make up the longest undeveloped shoreline in the conterminous states. A four-wheel-drive vehicle patrols the entire beach weekly, but only the first fourteen miles are kept clean of litter, and that at a cost of $3000 a mile per year. The rest of the coast lies blanketed under a mass of jetsam, including the rotting hulks of wrecked ships and enough light bulbs to illuminate New Orleans.

Once an imposing monument to bad seamanship, the 600-ton *Nicaragua* has rusted away till only a little dab of twisted iron remains of what Texas journalists insist on calling a "mystery ship." The story is that the vessel ran aground while running a load of illegal weapons to Mexican revolutionaries. Supposedly, Mexican agents slipped aboard the vessel and sabotaged the steering gear.

Actually, survivors of the first rescue and salvage party say the *Nicaragua* was only a banana boat—and an empty one, at that—when it went aground through sheer incompetence of the crew. But reports persist of great profits from arms traffic lying buried somewhere in the Padre Island sands. And Lafitte's corsairs supposedly buried a few chests of gold around there somewhere. One of the previous cattle ranchers buried the family treasure when he ran away from federal occupying troops during the Civil War. An immense Spanish treasure flotilla sank just off the coast. And so on.

Rarely does the park patrol fail to pull at least one treasure hunter's car out of the sand, and I realized that those camper trucks stuck to their hub caps that I saw on my plane flight were piloted by hunters after pirate gold.

Strangest of the objects found in all that jetsam was shown to me by Louis Rawalt at his Coastway Bait Stand just outside the park. Squatting with folded arms, a little stone figurine showed the flattened head, Negroid lips, and slanting Oriental eyes marking it as from the La Venta phase of the ancient Olmec Culture on the distant Mexican coast. Dated at about the time of Christ, the little man had wandered six hundred miles from his home—how and in whose pocket we could only guess.

At the southern end of Padre Island I was astonished to discover that the Rio Grande—the Big River—is a trickle making less of a delta than dozens of insignificant bayous and creeks in other coastal states. A major shrimping fleet ventures out from Brownsville, about twenty-five miles upstream, but that is the head of navigation, for generations of cattlemen and cotton farmers have considered that fresh water pouring into the Gulf is wasted, so they have trapped it behind dams to irrigate upstream lands.

From the air I had noted patterns of ancient oxbow cutoffs of the river, now dry arroyos called *resacas* in the local Spanish dialect (probably from a corruption of *rio seco* or dry river.) But the river is unlikely to form any new oxbows, for it is contained behind a levee on the American side.

Carlos Watson, an attorney with the typical Brownsville mix of Spanish and Anglo ancestry, took me on a tour of the farthest southern loop of the Rio Grande.

"With all the dams upstream, it's a rare high water that gets this far downstream," he said. "And then it benefits only Mexico, for they have had the good sense across the way not to keep the flood waters off their fields. That way they get all the fertilizing silt washed down from upstream farms, and we have to buy commercial fertilizer."

The river's delta does have its own beauty, nevertheless, with the same snowy sands and golden oats along the beach and the same salt marsh grasses inland that make the Gulf of Mexico shore the country's richest seafood nursery.

A Chicago developer is building a retirement and resort village within sight of the Rio Grande delta, but the natural beauty remains.

So I found it everywhere around the Gulf. Cities sprawl, pollution threatens life, litter defaces.

But much remains.

The tern nesting grounds of the Tortugas, the flowering exotics of the Keys, the teeming life of the mangrove jungle, the mud-flat nurseries around Florida's Big Bend, the dunes of the Panhandle and Alabama's shore, the new National Seashore off Mississippi, the incredibly fecund salt marsh nursery of Louisiana's coast, the bird havens along the Texas shoreline—they still shelter and feed an immense reserve of wildlife. Hundreds of miles of magnificent littoral have escaped the bulldozer and dragline.

The Gulf of Mexico shoreline is still worth saving.

Random growth of yucca, Atascosa Wildlife Refuge, Texas

Green Island, a bird sanctuary in the Laguna Madre, Texas ►

Marsh at Aransas Wildlife Refuge, winter home of whooping cranes

Sea oats, Padre Island National Seashore ►

Jellyfish, Padre Island

Sea oats, Padre Island

◄ St. Joseph Island, Texas

Wild Turkeys, Aransas Wildlife Refuge

Virginia white-tailed deer, Aransas Wildlife Refuge

White pelicans near Corpus Christi

South end of Padre Island ►

The delta of the Rio Grande with Mexico on the far shore